COMPREHENSIVE GUIDE ON AIR CONDITION AND ITS MAINTENANCES

Simple book on installing, servicing and mounting of HVAC device

Desmond F. White

Table of Contents

CHAPTER ONE

INTRODUCTION

Having a draft devices helps in facilitate welfare; a drafted air conditioning framework will have individual indoor regulators in various areas, alongside a focal control board for the whole framework.

Adaption's In Circulations

This permits the central air framework to keep up with independent temperatures among various regions of the house or building. Zone damper frameworks will commonly be mechanized to adapt to the singular temperatures set on each indoor regulator.

WHAT AND WHAT TO KNOW IN CENTRAL AIR EXAMINATION

In the event that you are new to these terms, the stockpile is where the hot or cold air emerges to control the temperature inside the room or region, and the return is where the air is moved once again into the central air framework from the adapted spaces. The return framework in central air assumes an urgent part in guaranteeing that the framework is working proficiently. Whenever air is maneuvered into the return channel, this makes a negative strain framework that will assist with bringing the molded air into the room, which is the reason you will frequently see supply and return focuses put on inverse sides of a room.

HOW TO PRESENT AND ADDRESS DEVICES

One more significant capability of the air conditioning return framework is air sifting. In the event that you have at any point taken out the grille from a return channel the metal covering with long, flat openings, you most likely found a channel set inside the bring conduit back. Nonetheless, on the off chance that you didn't find a channel here, you ought to introduce one since this will work on the presentation and life expectancy of your central air framework.

SPOTLESS AND CHANNELS

Central air return channels are utilized to eliminate dampness, residue, hair, and different particles that might be circling inside your home. By guaranteeing that spotless, separated air is gotten back to your central air framework, you can decrease the development of defilement inside your channels, warmer, and cooling frameworks. At the point when the internal parts of these frameworks are kept spotless, this will work on their presentation, lessen the requirement for support, and guarantee that they last longer.

EXPECTANCY OF A/C

This will likewise assist with working on the nature of the air that is being appropriated around your home. Return channels are genuinely modest, and they ought to be supplanted each 32 to 88 days, contingent upon what sort of channel you are utilizing. The life expectancy of your channel will likewise be impacted by different variables, for example, the region where you live and the nature of the channel.

CROSS CHECK AND SPEED

A decent guideline for an air channel is, in the event that it looks grimy, feel free to change it; Moves toward Introduce a High Speed air condition Framework. The initial step is to do heap estimation; Gives Way estimate structure that is intended to precisely give a room by room load for the high-speed framework. The underlying thought is the area of the fan curl unit. The fan curl unit might be situated in an unconditioned space.

SHIELDED AREA

Regions like a storage room, carport, or unfinished plumbing space are satisfactory the length of the area is shielded from the climate and neighborhood codes are met. The fan loop unit may likewise be situated in a molded space like a storm cellar, storeroom, or utility room. Two sorts of fan curl arrangement are offered level and vertical cupboards. One must likewise consider the area of the return air channel box and conduit. One more significant thought is the area of the 8-inch inside width plenum, air supply tubing, refrigerant lines, and condensate channel line.

CREATE LOWER AND UPPER

While introducing a fan loop unit in an upper room space, for example, over a roof, it is prescribed to introduce an optional channel dish as well as a float switch. Again really take a look at your nearby codes for this necessity. The unit might be introduced on a stage or suspended from a pre-collected suspended stage.

SELECT OPEN AND CUT

The stage fixed or suspended should be adequately close to the chose return air box opening for simplicity of interfacing the return air plenum conduit. Then, cut the return air opening. Select your definite area for the return air pipe.

LOOP AND OPEN

Try not to introduce the return air confine a lounge area, family room or kitchen except if the air conduit can be introduced with a 180-degree twist; when the opening is cut in view of the size and aspects of the return air open, the fan loop might be gone through the opening.

INCH UPWARD AND TWIST

The return air box is intended to fit between joists that are 16 crawls on focus with a 14-inch width. For wall applications, for example, while utilizing an upward unsupported fan curl unit, be certain the initial will consider a 90-degree twist in the return air conduit.

HAVE A PASS AND BOX

Next check for legitimate attack of your return air box; In any case, don't introduce the return air box until your establishment is finished in light of the fact that you might need to pass materials for your establishment through this opening.

SIZE LIGHT UP

You can develop a straightforward stage in light of the size of the fan loop unit. It is prescribed to utilize 2-by 8-inch stud blunder and at least 1/2-inch compressed wood for stage development. Continuously utilize gave segregation strips under the unit for best execution. For where the fan loop unit should be suspended, suspend the stage above by 1/4-inch strung bars.

OVERHAULING CURL

Never interface a strung pole straightforwardly to the fan curl bureau. It's not important to get the fan loop unit to the stage in light of the fact that the heaviness of the unit will stand firm on it in situation. Make sure to leave space for overhauling.

RING PLENUM AND ALIGN

Recollect that the stock plenum ring will be a few inches lower than the lower part of the fan curl unit. It will be important to move the fan curl past the stage edge to oblige the plenum ring.

MOUNT TRAP AND ASSOCIATE

The plenum ring joins to a square opening with four screws gave. Furthermore, make certain to append the plenum ring gasket to the fan curl bureau prior to mounting the inventory plenum ring. You are presently prepared to associate the condensate channel trap provided to the fan curl unit.

LINES AND INCH INTERFACE

Allude to the directions with the provided condensate channel trap. Continuously run the condensate channel line from the snare to a reasonable channel as per neighborhood codes.

Be certain that the channel line is pitched 1/4 inch for each foot. Never interface the condensate line to a shut channel framework. Continuously utilize an optional channel search for gold frameworks.

CHAPTER TWO

CONDUIT WIRE

GUIDE IN INTERFACE REFRIGERANT LINES

UNITS PHASE

Interface the refrigerant lines from the outside gathering unit to the indoor fan loop unit. Continuously follow the consolidating unit open air unit producer's directions for introducing, estimating, catching, charging, and utilizing channel dryers.

TUBING AND PIPE

While introducing the air circulation parts, it is dependably smart to do a framework format in view of your measuring boundaries.

ELBOWS AND RUN SUPPLY

The plenum pipe might be run in essentially any area that is available for the connection of the stockpile tubing. Plenum conduit establishment, all tees, elbows, and branch runs should be at least 18 creeps from the fan curl unit and some other tee, elbow or branch supply run. It is suggested that you keep the utilization of tees and elbows to a base to downplay the framework pressure drop. The plenum conduit comes in 6-foot lengths and might be sliced to wanted lengths relying upon your framework format.

EMBED AND COZY JOINT

While introducing the plenum to the unit, it is suggested you cut off the male finish of the primary area. Then embed the plenum pipe area into the getting collar on the fan curl unit. Make certain to push the conduit in close to frame a cozy joint.

TAPED JOINTS AND FOIL

Then level the sheet metal tabs against the plenum channel and supplement the level head pins. Then, wrap and tape set up the 6 all inclusive piece of foil face protection over the taped joint. Be certain the protection is pushed facing the fan loop unit.

PAD PHASE PINS

Keep on collecting the plenum channel ensuring the shiplap joints are cozy and taped safely. When your plenum pipe has been introduced, the plenum end cap will presently go on the finish of your plenum run. The plenum end cap contains a protected pad that is pushed up close against the conduit end. Hold the end cap immovably set up, then, at that point, embed the level head pins through the sheet metal tabs toward the finish of the cap.

OUTLETS ELIMINATOR

Cross tape the end cap safely with a circumferential fold over the end cap. It's consistently really smart to lash or support the plenum to stay away from plenum development during activity, which could make joints come free. You are presently prepared to start establishment of the room eliminators or outlets and the sound weakening tubing. Room eliminators and sound constricting tubing are given in the establishment units. Imprint the places where you might want to introduce the room eliminators.

DRILL AND POWER SOURCE

While denoting the area for the room eliminator, the focal point of the eliminator ought to be 5 crawls from a wall edge. At the point when introduced in a room-roof corner, the area ought to be 5 creeps from each wall edge. After you mark the area, drill a 1/8-inch opening for the power source.

FITTING AND CONNECTORS

Permit no less than 2 crawls of freedom overall around the 1/8-inch opening. After all clearances have been checked, utilize a 4-inch opening saw to cut an opening, utilizing the 1/8-inch opening as your pilot.

CREATE SWITCH AND FIT

Next gather the room eliminator to the sound constricting tubing by fitting the connectors together and winding until tight. In the event that the eliminator is being introduced in a story area, you should manufacture a little 1/4-by 1-1/2-inch-square screen and spot the screen between the eliminator faceplate and the tubing connector. Then push the free finish of the sound lessening tubing through the 4-inch opening until the switches on the eliminator fit properly. Focus the two spring cuts on a line lined up with the heading of the tubing so the faceplate doesn't pull away from the roof.

PRACTICAL INSTRUCTIONS ON SUPPLY TUBING

Curves and length

Fast connectors, plenum departures with gaskets, and offsetting openings are provided with the establishment unit.

Continuously keep away from sharp curves in the stock tubing and attenuator tubing. Supply tubing commonly comes in 100-foot lengths and might be sliced to wanted length.

Fold and neckline

The base length is 6 feet and the greatest is 30 feet with attenuator. Slice your stockpile tubing to the legitimate length. Then introduce the connector to the tubing by stringing it into the cut finish of the tubing. Fold the protection and Mylar scrim under the neckline of the connector. Fold tape over the connector to get the two pieces together. At last, introduce the connectors together by utilizing a turn activity until they are gotten. At the plenum area you have chosen, cut a 2-inch opening in the plenum with the opening shaper provided.

Pivot and edge foil

Be certain that your opening is at a 20-degree down point to wipe out a burden on the plenum. To cut the opening with the opening shaper, pivot the shaper this way and that, applying barely sufficient strain to drive the serrated edge into the foil and protection of the plenum. Eliminate the opening cut out from the plenum. Make certain there is no fold left that could impede the opening during activity. Then, place the plenum departure gasket around the opening.

Pincers placing

Then place the plenum departure connector into the opening in the plenum. Arrange the plenum departure to match the ebb and flow of the plenum channel. Embed the plenum clasp by hand each in turn. Utilizing pincers, adjust the clasp properly. When the plenum departure is set up, you're prepared to introduce adjusting holes if important. Then, introduce a connector into the excess open finish of the stockpile tubing following a similar methodology as in the past.

Rerun and channel

Then, at that point, interface the connector from the tubing to the plenum departure following the curving activity used to associate the connectors. You have now finished your inventory run. Whenever you have finished your stockpile run establishments, you can start to introduce your return air channel and return air box. Eliminate the return air grille and channel from the return air box get together. Embed the return air enclose to the return air opening you cut before and affix the casing with four screws through the openings gave on the long side of the crate.

Secure and clasp

Then, embed the return air grille into the case and secure with the four screws gave. Open the grille and addition the channel. Then, at that point, associate each finish of the return air adaptable pipe to the circular finish of the fan curl and the return air box associations with the clasp groups.

PRELIMINARY   ON INSTALLATIONS

Low and high voltage wiring

Refrigerant funneling

Electrical breakers

Ventilation work

Condenser cushion

Refrigerant levels

Wind current

Supply air temperature

Channel capability

APPARATUSES NEEDED

Responding saw

Vacuum siphon

Refrigerant scale

Cordless drill

Screwdrivers

Pipe wrenches and forceps

Cuts and shears

Multimeter and voltage analyzer

Wellbeing glasses

Gloves

Estimating tape

Flashlights

WHAT TO BEAR IN MIND

Wear gloves while taking care of sharp apparatuses or materials like ducting. Lift with your legs, not with your back and Try not to drink liquor at work.

BASIC TO CONSIDER AFTER INSTALLATIONS

SIZE

Accurately Size New Framework; Work out Warming and Cooling Burdens. Introducing an erroneously measured unit prompts untimely mileage, more limited life expectancy of the framework, expansion in energy bills and conflicting temperatures all through the home during various seasons. A certified expert ought to lead load estimations by get-together the right data prior to making a proposal, so in the event that you're doing estimations yourself, be pretty much as careful as could be expected.

CURRENT

Air conditioning frameworks are costly, so it's vital to ensure they're working appropriately. First check the ventilation work for any harm or trash that could be impeding wind current. Clear all impediments, the majority of which you'll find at surge and data registers frequently called sends and returns.

SPACE OCCUPY

Pick the area for your unit, whether you've picked an indoor or outside framework. You should do a careful site assessment then mark out utilizing splash paint or chalk where you believe the unit should go.

www.ingramcontent.com/pod-product-compliance
Lightning Source LLC
Chambersburg PA
CBHW060016300526
45794CB00003B/1200